Geography Zone: Landforms™

BEACHES

Emma Carlson Berne

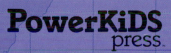
PowerKiDS
press™

New York

Published in 2008 by The Rosen Publishing Group, Inc.
29 East 21st Street, New York, NY 10010

First Edition

Editor: Joanne Randolph
Book Design and Photo Research: Julio Gil

Photo Credits: Cover, pp. 5, 7, 9, 11, 13, 15, 17, 19 Shutterstock.com; p. 21 © AFP/Getty Images.

Library of Congress Cataloging-in-Publication Data

Berne, Emma Carlson.
  Beaches / Emma Carlson Berne. — 1st ed.
      p. cm. — (Geography zone. Landforms)
  Includes bibliographical references and index.
  ISBN 978-1-4042-4205-0 (lib. bdg.)
  1. Beaches—Juvenile literature. I. Title.
  GB453.B47 2008
  551.45'7—dc22
                                                2007032304

Manufactured in the United States of America

Contents

Learning About Beaches

A beach is a place on a sea or ocean shore that has sand or rocks. Beaches can be made of fine sand or big rocks. Beaches can also be made of pieces of shells. Some beaches are very short, just a few yards (m) long. Others are as long as 100 miles (161 km).

Beaches are always changing. They look different in different seasons. Waves push the sand or rocks to different places. The movement of the water can dig out holes. The wind blows along the sand, moving it around. Let's learn more about beaches!

There are beaches on coasts around the world.
This sandy beach is in the Netherlands.

Rivers and Beaches

What did the river say to the beach? I would like to make a **deposit**! Rivers and waves work together to make beaches. Rivers carry sand and rocks. When the river meets the ocean, it deposits the rocks and sand on the ocean bottom. Then waves push the sand and rocks back onto the shore, making the beach.

In the summer, many beaches are wide and rise gently up from the water. In the winter, beaches can become steep and narrow. Strong winds and storms make stronger waves. The powerful waves toss sand onto the shore and then take it away.

The place where a river joins the ocean is called its mouth. You can see where some of the sand the river has carried has been deposited here.

Parts of a Beach

A beach has two basic parts. The sand closest to the water is called the foreshore. This part gets wet from waves. The sand further away from the water is called the backshore. The backshore sand does not get wet, even at high tide.

The line of sand between the backshore and the foreshore is called the berm. The berm looks like a long mound of sand, sometimes with a flat top.

At the very back of the beach, there might be dunes. Dunes are hills of sand that have grasses growing on them. Many birds like to live in dunes.

Here you can see the wet sand, the dry sand, and the grassy dunes that make up the different parts of a beach.

Sand, Sand Everywhere

Many beaches are made of sand. Sand is tiny bits of rock or **mineral**. Sometimes, beach sand is made of a mineral called quartz. Beach sand can also be made of the bodies of tiny **organisms** that live in the ocean. Some sand is even made from cooled-off **lava**.

Sandy beaches can be different colors. Lava sand is black. Sand can also be yellow, white, or even pink and green.

Most **tropical** beaches have white sand. On the island of Bali, in Indonesia, there are miles (km) of white sand beaches.

This black sand beach is in Iceland. The sand is made of basalt, a kind of rock that is made when lava cools and hardens.

Some beaches are made of rocks, **boulders**, or **pebbles**, instead of sand. Rocky beaches are usually narrow and steep.

The ocean near rocky beaches has powerful waves that can pick up and move the rocks and pebbles. Boulders sometimes crash down onto the shore from cliffs. They are too heavy for the waves to move. Over time, the waves will wear the boulders away until they become just like other rocks on the shore.

In Maine, the beaches are made of rocks and boulders. Seals sometimes crawl out of the ocean and lie in the sun on the boulders.

This is a rocky beach in Maine. Over time, the waves will wear away the rocks until they become smaller and at last become sand.

Shells and Coral

Most beaches are sandy or rocky. Shell and **coral** beaches are more unusual. These beaches are made of crushed shell pieces or bits of dead coral. The coral commonly washes ashore from coral reefs off the coast. Coral reefs are underwater hills of living and dead coral. Sometimes, fish bones and shark teeth are mixed in. On one shell beach in South Carolina, people have even found pieces of dinosaur bone!

Shell and coral beaches are usually white or pink. They are usually found in tropical places, but the coasts of Florida, the Carolinas, and Vancouver, Canada, all have some shell beaches.

These shells are on a coral beach on the coast of the Red Sea, between Egypt and Saudi Arabia. Lots of coral grows in the Red Sea's warm waters.

Beaches Are Homes

Many animals live on beaches. By the water's edge, clams bury themselves in the sand as waves wash over them. Flies and other biting bugs live farther up on the beach, away from the water.

Waterbirds like sandpipers and plovers run along the shore with their feet in the water. They peck up tiny bugs and shellfish hidden in the sand.

Sometimes, sea turtles crawl from the waves at night. By the light of the Moon, they lay their eggs in nests built in the sand. Then they crawl back into the ocean.

This hermit crab sits on a piece of coral on the beach it calls home.
As a hermit crab grows, it will leave its old shell for a larger one.

A Hawaiian Beach

Waikiki Beach is one of the most famous beaches in the world. It is in Hawaii, near the city of Honolulu. Every year, thousands of people visit Waikiki and enjoy its white sand.

The sand on Waikiki Beach is not washed up there by the waves, as it is on other beaches. The sand is brought to the beach in trucks, dumped out, and spread around. This is because the beach at Waikiki is always being **eroded** by the waves. People still want to use the beach, though. They do not want it to be washed away. They have to bring sand in or the beach will disappear.

Many hotels line Waikiki Beach, since it is such a well-liked spot for people to visit. Hundreds of thousands of people have come to Waikiki.

Surfing is a popular sport at Waikiki Beach. People also dive, paddle sea **kayaks**, and swim in the warm blue water around the beach.

Huge hotels line the shore. **Tourists** come from all over the world to stay in these hotels and visit the beach.

The weather is warm in Hawaii most of the year. All through the seasons, people lay on blankets on Waikiki Beach. They take walks up and down the beach with their feet in the water. Sometimes, people even get married on the beach.

These two surfers wax their surfboards before hitting the waves off Waikiki Beach. Many people come to this beach just to surf its warm waters.

Keeping Beaches Safe

All over the world, people love to live near beaches and visit them. Whether they are rocky, sandy, or made of shells, beaches are beautiful places.

People can hurt beaches, though. People sometimes litter on beaches. The trash **pollutes** the shore and the ocean. People have also built dams or destroyed dunes. These things are causing beaches to disappear.

Many countries and cities try to keep their beaches safe. They ask people not to litter or touch nests, so that both animals and people can share the beach. How can you help?

Glossary

boulders (BOHL-derz) Very large rocks.

coral (KOR-ul) Hard matter made up of the bones of tiny sea animals.

deposit (dih-PAH-zut) Something that is left behind.

eroded (ih-ROHD-ed) Worn away slowly.

kayaks (KY-aks) Small, light boats steered by paddling.

lava (LAH-vuh) Hot, melted rock that comes out of a volcano. A volcano is an opening in a planet that sometimes shoots up lava.

mineral (MIN-rul) A natural thing that is not an animal, a plant, or another living thing.

organisms (OR-guh-nih-zumz) Living beings made of dependent parts.

pebbles (PEH-belz) Small, rounded rocks.

pollutes (puh-LOOTS) Hurts with harmful matter.

surfing (SURF-ing) Using a board to ride ocean waves.

tourists (TUR-ists) People visiting a place where they do not live.

tropical (TRAH-puh-kul) Warm year-round.

Index

B
boulder(s), 12

D
deposit, 6

K
kayaks, 20

L
lava, 10

M
mineral, 10

O
ocean, 6, 10, 12, 16, 22

organisms, 10

P
pebbles, 12

R
rock(s), 4, 6, 10, 12

S
sand, 4, 6, 8, 10, 12, 16, 18

shells, 4, 22

shore, 4, 6, 12, 16, 20, 22

surfing, 20

T
trash, 22

W
Waikiki Beach, Hawaii, 18, 20

water, 4, 6, 8, 16, 20

waves, 4, 6, 8, 12, 16, 18

Web Sites

Due to the changing nature of Internet links, PowerKids Press has developed an online list of Web sites related to the subject of this book. This site is updated regularly. Please use this link to access the list:
www.powerkidslinks.com/gzone/beach/